ABOUT DARK ENERGY AND DARK MATTER

A SIMPLE EXPLANATION

Dante Barbis Ayres

1

Foreword

I want to thank my friends M. Augusto Blacker and Christian Samanamu, for reviewing the originals of this document and for suggesting some valuable corrections.

Peru, 2015

CONTENT

SCOPE OF THIS PAPER

This paper tries to provide a simple explanation of dark energy and dark matter. And, what either might be.

In the process, we will try to explain the following:

The types of particles that constitute dark energy, as well as, and the kind of particle that constitutes dark matter.

The connection that exists between the most basic dark energy particles and the invariance of the speed of light.

The connection that exists between dark matter particles and the gravitational force. It concludes on the possibility that both types of particles: dark matter and gravitons could be the same.

Finally, and based on the properties of the particles described, we try to give an explanation about the change on the expansion velocity of the universe.

THE ETHER: AN OLD THEORY

The idea of the ether arises from the principles of classic mechanics by which we conceive that the transmission of light through space requires a medium (ether) so that it (light) can move through or within it.

Many years ago, trying to locate the absolute system (stationary ether), several experiments such as those of Trouton and Noble[1] were carried out.

Also, a well-known experiment was carried out for Michelson and Morley.

It was then proposed that, if the ether existed, an observer on earth could detect the "ether wind" and, that its speed would be the orbital velocity of the earth. For this, a shift on the interference pattern should be detected with a Michelson's optical interferometer, when a ray of light passes through the instrument.

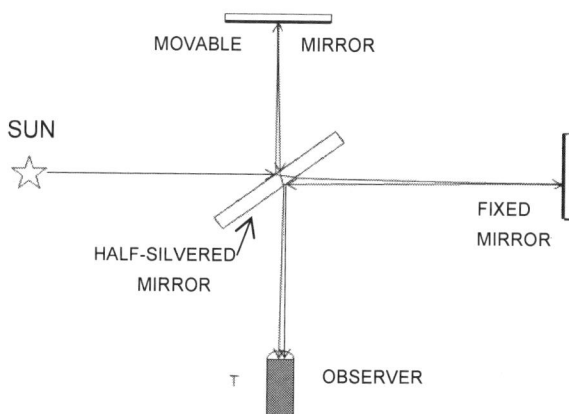

---

[1] PANOSFKY, Wolfgang K.H. and PHILLIPS, Melba. Classical Electricity and Magnetism. Pps. 274-275. Addison Wesley Publishing Co. Inc. Reading, MASSACHUSETTS. 1962.

Light from sun is split in two beams by a half-silvered mirror and reflected by two others mirrors placed at 90° one each other. The two light beams return through the half-silvered mirror to the telescope at "T" where, the interference pattern must to shift when the interferometer is rotated through 90°.

This famous optical experiment was first performed by A. Michelson in 1881, when he was trying to discover the movement of earth through space.[2] No such shift was observed.

The initial negative result was surprising and disillusioning. The experiment was repeated and a more precise effort performed by D. C. Miller by using a Michelson and Morley's instrument with larger optical trajectories but still provided a negative result.

Other hypotheses were proposed to save the idea of the existence of "the ether", and to explain the negative results of the experiments with the optical interferometer, such as the "contraction of bodies", by Lorentz-Fitzgerald, and that of the "attraction of ether"[3].

The latter assumption indicated that the ether, in contact with bodies with a finite mass, adhered to them and that therefore, this portion of fixed ether had zero velocity with regard to these bodies. In essence, that each body carried with it its own, local ether.

That was the way the negative result of the Michelson-Morley experiment was overcome. However, the hypothesis of the attraction of ether was also discarded when well-known phenomena were taken into account; such as those of the stellar aberration and Fizeau's convection coefficient.[4]

---

[2] PANOSFKY, Wolfgang K.H. and PHILLIPS, Melba. Classical Electricity and Magnetism. Pps. 275-277. Addison Wesley Publishing Co. Inc. Reading, MASSACHUSETTS. 1962.

[3] PANOSFKY, Wolfgang K.H. and PHILLIPS, Melba. Classical Electricity and Magnetism. Pps. 2787-280. Addison Wesley Publishing Co. Inc. Reading, MASSACHUSETTS. 1962.

[4] PANOSFKY, Wolfgang K.H. and PHILLIPS, Melba. Classical Electricity and

No objection exists up till now to Albert Einstein's special theory of relativity, one of whose points is the invariance of the speed of light. This means that the speed of light in the vacuum has the same value = c in all inertial systems and also, the proven fact that the propagation of a luminous signal is independent of its source.

The above, is precisely one of my concerns: that everybody accepts the invariance of the speed of light but, until now, there are no reports trying to explain why light or electromagnetic waves have precisely that speed or what is that physically determines it.

On the other hand, we know that vacuum is not empty. It is filled with dark matter and dark energy. Therefore, a portion of that dark energy may be the so much sought-after "ether" which allows or serves as a medium for the transmission of light.

Magnetism. Pp.280. Addison Wesley Publishing Co. Inc. Reading, MASSACHUSETTS. 1962.

WHAT KIND OF PARTICLES ARE THERE IN THE VACUUM?

Based on the principle of the invariance of the speed of light, I have developed a new theory about the existence of a common environment.

To start, I reject the long-held presumptions of many years ago about the nature of the ether such as the one that says that "the ether" is a system at rest, or that it is formed by particles that are normally at rest (stationary ether) with a zero mass and that could eventually be attracted. Remember that the properties attributed to this ether were somewhat strange: zero density and perfect transparence[5].

These ultra–tiny particles or "basic particles" that, as it was thought, made-up the ether would permit the transmission of any signal. But, this would mean however, that atomic particles that produce a disturbance, or any signal to be transmitted, should create that signal with a speed of propagation equal to c = 299,792,500 m/sec. It was never thought that this was not strictly true because an infinite amount of energy would be needed to transmit the signal towards the infinite or we would need to "invent" particles with zero mass in charge of transmitting signal to the infinity.

In accordance with the original conception, since "the ether" only serves as a medium, the signal would have to be created or be delivered to "the ether", with a speed such as c. If not, how could it be possible that these particles – standing at rest– could transmit a signal at the speed of light?

Furthermore, not always a signal is fed to a medium at the speed of light. We know that the atomic particles that interact amongst themselves or that change their velocity emit radiation (for example: an electron when passing from

---

[5] ROBERT, Resnick, Introduction to Special Relativity, Pp. 16, John Wiley & Sons, Inc., Spanish version, Editorial Limusa, Mexico, 1977.

one orbit to another because of a deceleration emit photons), and that radiation propagates at the velocity of light even though the atomic particles that caused the radiation were moving at a speed below that of c. Another example is about the electrons of a cathode ray moving at a speed below that of c that impacts behind a TV screen but the image waves that the screen emits travel at the speed of c.

Therefore, we must conclude that in all these different cases, the environment (the medium) receives a disturbance, and then, it (the medium) propagates that disturbance at a speed = c. This, without requiring that the atomic particles that created the signal or disturbance, did so moving at speeds slower than of the velocity of light. Hence, this point out the fact that, the medium has the capacity –enough energy and intrinsic speed– to accept a signal and transmit it at the speed of light.

I must also say that I am not in agreement with Max Born, who emphasized the fact that: "The elastic properties of matter were deduced better and better from the electromagnetic forces, and that it would be illogical, in turn, to attempt to explain electromagnetic phenomena in accordance with the elastic properties of some hypothetical environment".[6]

What in my opinion, really occurs in the environment, or the medium, or the ether (if you wish) –where the all-visible matter exists– is the following:

The environment (the vacuum) is filled with what I call "Basic Particles". We know that vacuum is not really empty because it contains Dark Matter, and also, plenty of Dark Energy.

Let me now touch on the question of how these most basic energy particles could be.

---

[6] RESNICK, Robert. Introduction to the Special Relativity (Introducción a la Teoría Especial de la Relatividad). pp. 33. Editorial Limusa, México. 1977.

The space, the universe, is full of several kinds of "Basic Particles" smaller than the atomic particles we know.

The simplest of these ultra-tiny particles are moving (they are not at rest) and colliding into each other in a perpetual movement similar to molecules in a gas. This we can call a "vibrating environment".

The most important part of this theory is that these simplest basic particles –SBP– have their own speed. And, that it is approximately 41% faster than that of the speed of light as we will see later. Further, if we create a disturbance in any part of that medium in order to make a wave, and if we would take a given direction and measure the average speed of the front wave that was created, we would find that the average speed of that wave is = c.

The fact that all the space, including the space inside the atoms, is filled with these basic particles, and that they are moving at a speed of 1.41c in all directions (the vibrating environment), makes the following conclusions possible:

a) The relative speed of any body in relation to this environment cannot be measured because this vibrating environment is not at rest with regard (in relation) to any other system or body in the Universe. The particles that make up the vibrating environment are moving in all directions at the same time and its average velocity is almost 41% more than c. That is why Michelson-Morley's experiment could not determine the existence of an inertial system of reference; investigators, starting with them, assumed that it was at rest or in a uniform movement in a given direction. This environment however, is moving in all directions at the same time!

b) This particles cannot be dragged nor can they be detected as we will see later, and

c) Any perturbation or signal at a given point, causes the SBP particles being altered in their movement, to convey this alteration to their neighboring basic particles, and these in turn, to the ones next to them, transmitting that perturbation until infinity, and at a velocity = c .

Therefore, once this environment receives a signal (perturbation), it propagates that signal at the speed which is inherent to the environment, and in all directions; so, the propagated speed will be always c.

The explanation for the invariance of the speed of light is specifically based on the nature of a vibrating environment formed by mobile particles. Mobile particles that are constantly moving and colliding into each other at a velocity greater than c.

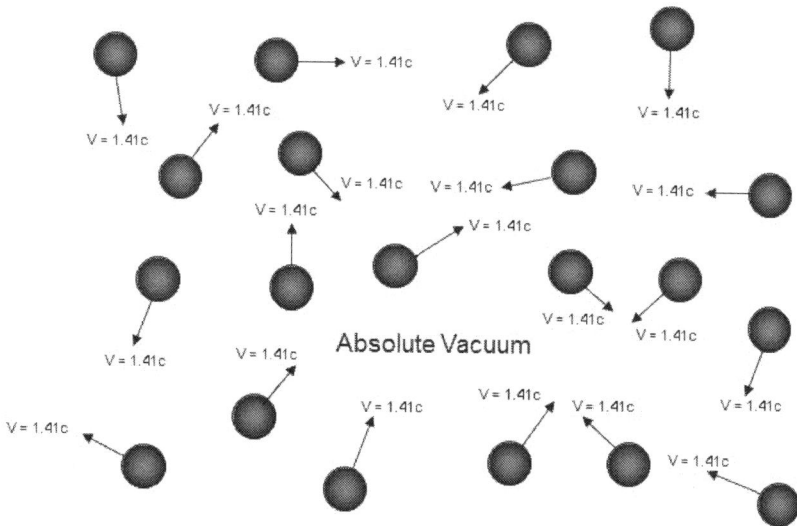

Simple Basic Particles - SBP

The simple basic particles –SBP– have mass and their mass is invariable. Even though these particles move at a higher speed than that of light, their mass has not grown until it becomes infinite nor will it be zero at rest (at the precise instant in which two of them collide front to front).

According to my theory, the variance of mass is given for all atomic particles: baryonic particles, or molecules, or bodies that move within the vibrating environment but, not for the basic particles that make up the vibrating environment. These basic particles may continue to collide with each other for thousands of millions of years and their average velocity will continue being the same. They are moving in the absolute vacuum and there is not any kind of friction that could put a brake on their movement. These particles are thousands of times smaller than an electron; they are compact, have no internal parts, they are simple spheres of maximum hardness in the universe.

I call these particles the "Simple Basic Particle", or "SBP".

Although there are many differences between the propagation of electromagnetic waves in this kind of environment and sound waves in a gas, there are also certain similarities. From a mechanical point of view it is important to note the correlation between the average velocity of the molecules of a gas (V), and the sound velocity propagation within the gas (S).

If we analyze the following table (I made) we can reach certain conclusions:

SOUND VELOCITY IN SOME GASSES

(At 20 °C of temperature, at sea level (1 atm))

GAS	Average Velocity V (m/sec)	Sound Velocity S (m/sec)	Velocity Factor (**) V/S
Hydrogen	1,840.00	1,269.50	1.45
Helium	1,308.36	902.30	1.45
Water vapor (*)	615.38	410.00	1.50
Nitrogen	493.72	339.30	1.46
Air	486.00	331.45	1.47
Oxygen	461.60	317.20	1.46
CO2	392.28	260.00	1.51

(*) The water vapor temperature is considered at 100 °C.

(**) The differences in the velocity factor between one gas an another are due to the fact that some molecules are mono-atomic and other polyatomic and therefore the elasticity of their impacts is different.

The sound velocity is less than the molecules' velocity because the molecules of the gas –that can transmit the sound– are moving in all possible directions and not precisely in the referent direction in which we wish to measure the speed of sound. It is totally improbable that all the gas molecules that make up the front wave of the sound move in a single direction in a given moment and that the collisions with the following molecule be perfectly frontal and so on. If this were possible, then the sound velocity in hydrogen, for example, would be 45% faster.

What happens is that the front wave is formed at a given moment by millions of molecules that move each in a different direction and the average of the projection of all the velocity vectors at a referent direction is precisely the speed of sound.

If V is the average velocity of the molecules of a gas, the average of the projections of these velocities in a given direction (taking into consideration only those molecules on a side of the perpendicular plane to the referent direction) must be:

$$V \cdot \cos \Phi \cdot EF$$

The average of the Φ angle is 45°. EF is the Elasticity Factor of collision of the gas molecules.

Therefore the sound velocity S will be:

$$S = V \cdot 0.70711 \cdot EF$$

The same thing happens with the transmission of light. The basic particles that at a given moment are making up the front wave of an electromagnetic signal are moving in all directions (at velocity of approximately 1,41c) and, we must project each basic particle's velocity vector in a single given direction. Each particle's velocity vector must be multiplied by Cos Φ. The elasticity factor EF must be 1.

Lots of basic particles take part in the formation of a specific electromagnetic wave, each with a given direction at a given instant. A specific particle could be precisely in a referent direction (Φ = 0°) at the instant of impact with the following particle but, this second particle, due to the direction it has will rebound in an angle between 0° and 90° (with regard to the referent direction), provided that as a product of the impact it will contribute towards the propagation of the wave. In other words, it will rebound on the same side of a perpendicular plane to the direction of the wave propagation.

Common sense tells us that we can consider the average of Φ (from 0° to 90°) as being equal to 45°. So, the factor between the basic particles average velocity V_{SBP} and the electromagnetic wave velocity must be

$$c = V_{SBP} \cdot Cos\ Φ$$

$$V_{SBP} = c/Cos\ 45° = c/0.70711 = 1.41c$$

Therefore, the average velocity of the basic particles that make up the vibrating environment must be 1.41 c.

Just imagine the energy a SBP has, moving at V_{SBP} = 1.41 c. Even the small mass the SBP has, its mass could be one thousandth of the electron mass, but there are billions and billions of these basic particles for each electron in the Universe.

But the most important point is that this medium has the capacity –enough energy and intrinsic speed– to accept a signal and transmit it at the speed of light.

At the precise moment in time of the Big Bang this powerful environment probably delivered part of its energy in order to build the different atomic particles we currently know; then, and after a while, the Universe reached

14

equilibrium, and the different atomic particles got their own energy –it of course, could change depending on the energy interchange with other atomic particles– but, certainly sometime after the Big Bang all basic particles' velocity entered a stable state, and have maintained their energy ever since; and, also their present velocity.

There are also two others "Basic Particles" filling the vacuum space as we shall see in the next section.

MORE BASIC PARTICLES

To complete the idea of dark energy and dark matter we have to review two other particles that fill vacuum space. These particles should represent an intermediate phase between the "SBP" and other known atomic particles like electrons and quarks.

The first one, which I call "BAP", is spherical; it looks as having a dense constitution with two or three cavities on its surface. The size of these particles could be about 20 to 30 times larger than the simple basic particles. The cavities on their surface appear to have been made by other particles that impacted them such as large meteors falling obliquely on the surface of the earth. The effect was similar to a cavity made by a small teaspoon which has taken more material from one side than from the other leaving a hole in the form of a tear.

These particles rotate because the "SBP's" beats against at each particle's cavities oriented in the same direction. And, the received energy is transformed into a rotation of the "BAP".

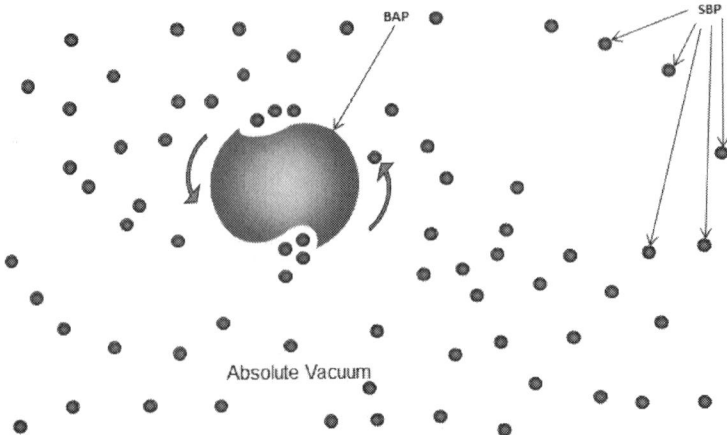

Basic Attraction Particles - BAP

16

In the end, there is a balance between the energy received from the SBP and the energy produced around it in the form of waves.

The waves produced in the cavities of these particles are like waves of vacuum. As the BAP rotates, their cavities are moving forward in a given direction and the simple basic particles that reached into every cavity are going to move in that direction. This effect leaves an empty space behind them which is replaced by other basic particles, generating a wave of "vacuum" that spreads radially at the speed of light. These particles form an attractive field around them that spreads to infinity.

I call these particles the "Basic Attraction Particle" or "BAP", and they should constitute the basic element in the buildup of what is known as dark mass. They could also be the "Gravitons", responsible for creating the gravitational fields of atomic particles.

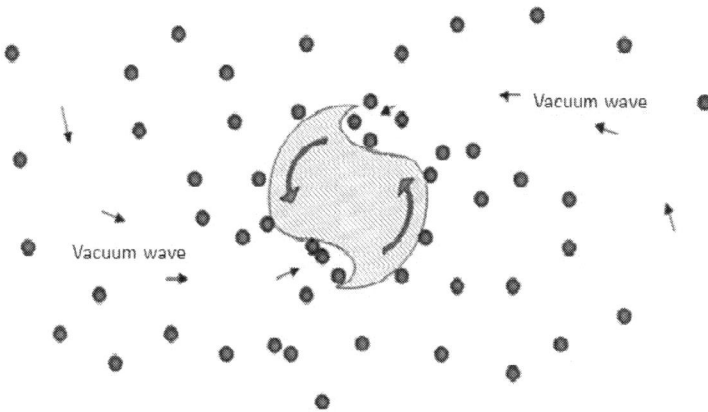

Vacuum wave

Vacuum wave

The other type of particle that fills vacuum space has also a circular shape; it is of equal size to the BAP.

These particles are not exactly spherical. They resemble a thick sheet, with warped ends in the form of a letter S. These particles, that I call "BRP", are also spinning rapidly, like an anemometer, because the SBPs colliding with the concave part of the BRP deliver more energy than the SBP crashing on the convex part of these BRP.

This new particle, because its spins, gets a balance between the energy received from the SBP and the energy it delivers to its medium, in the form of a pressure wave which is generated by the two convex parts of the particle. This pressure wave is propagated at the speed of light in a radial direction, rejecting all the particles around they reach; thus, creating a repulsive field that spreads to infinity.

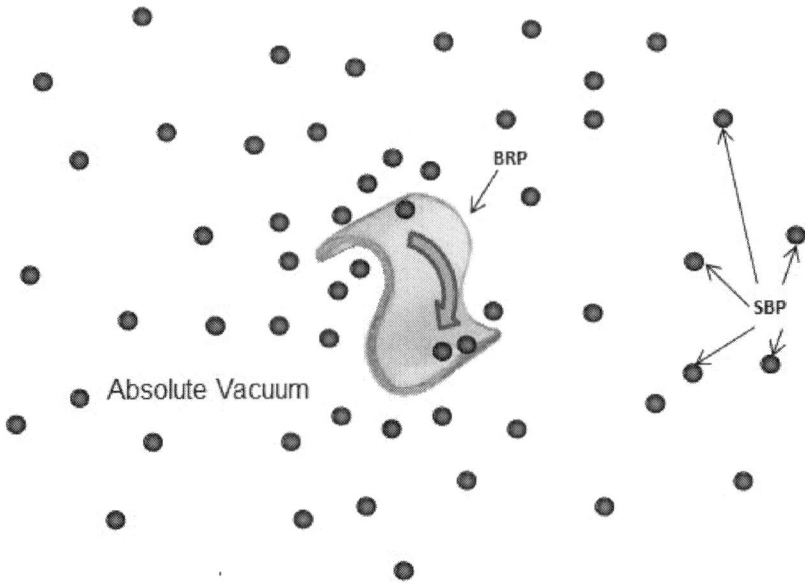

Basic Repulsion Particles - BRP

This type of particle would be a "Basic Repulsion Particle" hence, its name "BRP".

Given their size they cannot be detected, and because of their nature and way they carry energy, they are responsible for the expansion of the Universe. Hence, BRPs should therefore, be part of Dark Energy.

In fact, BRPs create a combined wave; where its most important characteristic is its repulsive nature or "pressure wave", generated by the convex part of the BRP but, followed by a weak vacuum wave produced by its concave side.

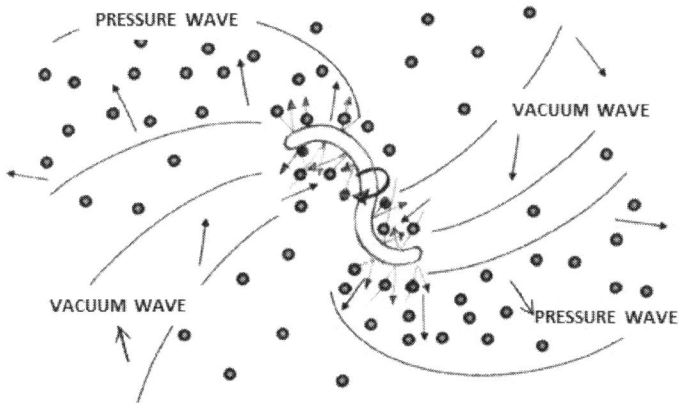

In summary, I conceive of a total of three types of basic particles.

The simplest and small of them –the "SBP"– fill all the space of the Universe that we believe is empty, and its energy is much greater than the energy of all the visible matter.   Its mass multiplied by the square of its velocity, which is equal to 1.41c, multiplied by their number (billions of times greater than all the particles of visible matter), should be an impressive figure. These particles do not interact directly with the atomic particles that we know but, they do it through the "BRP" and the "BAP".

It is possible that, at the time of the Big Bang those SBP probably moved more quickly, perhaps 2 times the velocity of light, and that therefore, they had more energy. However, they must have given part of that energy to the basic attraction particles –"BAP"– or dark matter, as well as, to the basic repulsion particle –"BRP"– or Dark Energy.

When all of these particles: the "BAP" and the "BRP", reached the rotation speed they have today a balance was established between the energies of these three basic types of particles. The "SBP" with its actual speed equal to 1.41c, and thus, defining the speed of light we know.

"BAP" and "BRP" particles should interact with other bigger particles such as the electron and quarks. These two particles –"BAP" and "BRP"– transmit in turn to the "SBP", all the signals they receive from electrons, neutrons and protons, whether electrical and / or magnetic.

In other words, these signals are transmitted by "SBP" at the speed of light to other "BAP" and "BRP" which could be at the proximity of other electrons, protons, and neutrons. And, because of that, the transmission at the speed of light of electromagnetic signals is made possible throughout the Universe.

I also think that electrons and quarks should have "BAP" inside them, like the gravitons that we cannot detect but, which allow a particle to attract another in proportion to its size. That is, in proportion to the number of "BAP" particles that can fit inside it. "BAP" particles remain self-contained within the electrons and quarks.

"BRP" particles on the other hand, may also enter in an electron or a quark but, cannot stay inside because its characteristics force these particles to reject everything and, in the end, they would end up expelling each other.

Therefore, the "BAP" would be the only one of these basic particles that could be self-contained within the known particles such as electrons or quarks and therefore, give them the gravitational characteristics to these particles, as well as, the mass properties we know.

Let me now speculate a little about the signals that these particles create around them. The attraction or repulsion waves created by "BAP" and "BRP" particles around them do not appear to be of a continuous nature but more of a digital nature. This because other particles feel these forces coming as pulses.

We may therefore say that, each "BAP" or "BRP" particle builds up around them an attractive force field or a repulsive force field. As indicated, those force fields appear to be not of a continuous nature but of a digital nature. Any particle inside those fields is going to be under a pulsating type force which is either attractive or repulsive, as seen in the following figures.

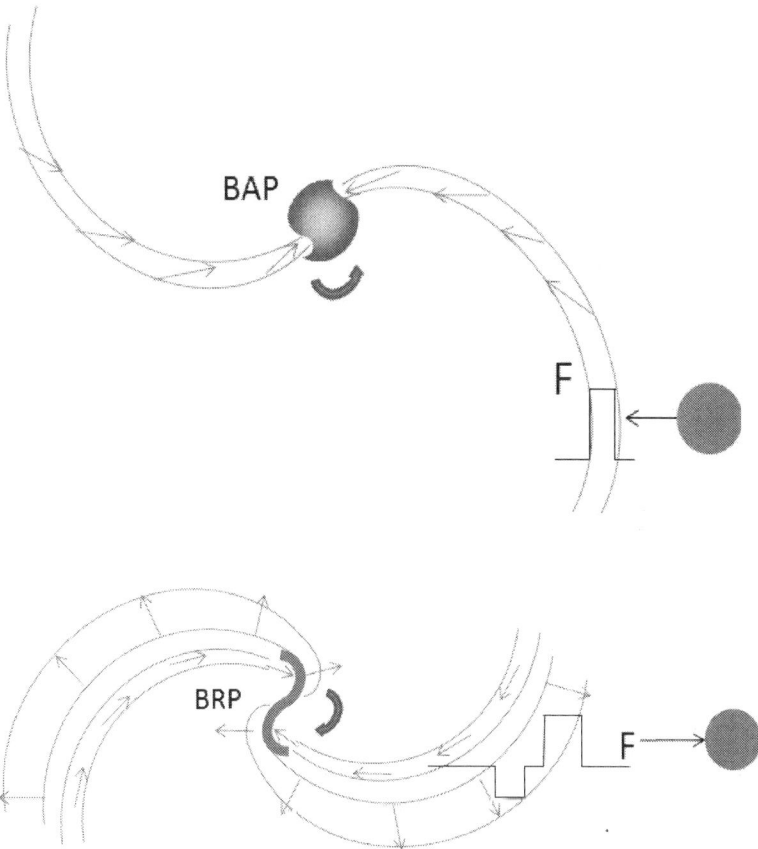

Because of the above, it is possible to confidently say that our reality is digital or Quantic –if you so wish– but not continuous.

WHY BASIC PARTICLES CAN NOT BE DRAGGED OR DETECTED?

We have already said that the basic particles that make up this new conception of the environment are of a dimension and mass that is far smaller than the smallest known atomic particles, such as the electron.

These basic particles fill all space, and also of course, interatomic spaces. They are responsible for transmitting electromagnetic signals between the atomic nucleus and its electrons, and between the electrons of an atom and those of the neighboring atoms which is part of a molecule.

The basic particles that make up this vibrating environment are the basis of the existence of gravitational and electromagnetic fields.

Normally, the SBP particles are moving around a fixed point in space. Their average journey is very short even though they do it at a higher speed than that of light since they quickly find other basic particles with which they collide and change their trajectory.

This continuous movement ruled by chance in the vicinity of atomic particles is greatly modified however, since these basic particles build up the environment which supports the electromagnetic interactions which allow atomic particles to remain linked within the nucleus, and electrons to remain in their orbits. The electric fields and the magnetic fields are the resulting configurations of the basic particle's movement and directions. The intensity of these fields result in reality, a greater density of all the basic particles (SBP, BAP, and BRP) at a given point, and at given moment.

Therefore, these basic particles (the SBP) that make up the environment are normally around or in the vicinity of a specific place in space. When a body such as the earth draws near, each of the basic particles will go through the atoms of the molecules of the Earth, varying the direction of their movement in order to make up the electromagnetic field of each atom that passes at every moment, and changing their trajectory billions and billions of times but, remaining around the same place once the earth has passed. This environment is not dragged.

The same phenomenon occurs with all the "Basic Particles" in free space. In interstellar space they are in movement by chance; movement which is only modified, by the transmission of light coming from the stars, as well as, by the gravitational fields coming from stellar bodies.

It is unnecessary for these basic particles to move beyond their current positions, even if they have to transmit light from one galaxy to another. They simply transmit the signal to the following basic particles and then, remain colliding with other basic particles. That is how an electromagnetic or gravitational signal can displace itself towards infinity without needing a "photon with no mass" or the injection of additional energy. Signals, after a first instant, become independent of their source.

It is also easy to imagine why no atomic particle and therefore no molecule or body can move at a higher speed than light.

This is so because the existence of atoms and molecules is based on the vibrating environment whose constituent particles have c as the limiting velocity in a given direction. Even more important is the fact that for an electron or an atomic particle to accelerate, some electromagnetic fields must act upon it. However, since the components of this electromagnetic field move at a maximum velocity that is equal to c, no matter what the field intensity is, its effect on a given particle will not be to push it to move at a greater speed than that of c.

If the atoms that make a body try to move at a speed that is higher than that of light, the basic particles would not have the capacity of transmitting the internal electromagnetic fields of each atom nor the electromagnetic fields between them; at least, in the direction of the movement. Therefore, the distance between atoms and molecules would be shortened in this specific direction as they drew nearer the speed of light, and they would disintegrate at the moment they reach this velocity.

It should also be noted that the elements and equipment we have in a laboratory to detect atomic particles deal with ionizations or traces left by these atomic particles as a result of collisions with other particles or by their interaction with electromagnetic fields. Atomic particles have a "field of matter" that is nothing else than a specific configuration of the basic particles due to the effect of the atomic particle's spin, as well as, the other known qualities of each atomic particle such as precession, mass and electric charge.

The "SBP", the "BAP", or the "BRP" particles cannot be detected principally because of their size. The instruments we have are too gross to be able to detect them. These particles do not have any electric charge; they have mass but do not have gravitons particles inside.

For example, the "SBP's" are simple spheres of dense material that are not affected by any type of known field.

DARK MATTER, DARK ENERGY, AND THE EXPANSION OF THE UNIVERSE

Now let's talk on a cosmological level in order to explain how these particles have influenced the expansion of the Universe.

While measuring the red shift of light from stars and galaxies, Edwin Powell Hubble, found that the Universe is presently expanding. The most recent observations allow us to say that the expansion of the Universe was in a decelerating mode since the Big Bang occurred until a few billion years ago, when the expansion begun to accelerate again. My assumption is that dark energy is causing the actual accelerating expansion of the Universe.

There is also evidence about a kind of dark matter that explains the motions of galaxies, since the masses of matter they contain are not sufficient to explain their attraction forces.

Basic attraction particles –"BAP"– could be the dark mass causing this phenomenon given their attractive nature. On the other hand, the other particles mentioned in this essay, the SBP and the BRP, could be the dark energy that is acting as described above because their repulsive nature.

One other question that comes up is what proportion of each of these particles exists in the Universe?

The proportions between the quantities of these particles are given by the findings of the Wilkinson Microwave Anisotropy Probe WMAP. The Wilkinson Microwave Anisotropy Probe (WMAP) is a NASA Explorer mission that launched June 2001 to make fundamental measurements of cosmology.[7]

According to that information, all matter corresponding to all galaxies, stars, planets, and living beings; in short, everything we see and is made of

---

[7] NASA home page, http://map.gsfc.nasa.gov/

electrons, protons, neutrons and other particles detectable in a laboratory, amounts to only 4.6% of the total.

The other 95.4% is dark matter and dark energy, plus a small percentage of neutrinos. Dark matter is approximately 24% of all matter in the Universe, and the rest, amounting to 71.4% must be dark energy.

We can therefore, assign 24% of the total mass of the Universe to basic attractive particles BAP and, the other 71.4% would be composed of simple basic particles SBP and basic repulsive particles BRP.

How can we to distribute these 71.4%?

Usually the simplest explanations are better in order to explain a new theory (Occam's razor).

I think that using a simple proportional ratio, basic repulsive particles "BRP" should have the same mass than that of the basic attractive particles "BAP": 24%. And, if we subtract that amount from 71.4%, we conclude that "simple basic particles" or "SBP" should represent 47.4% of the total mass of the Universe.

Using the above data we can then, explain the mechanics of the expanding Universe.

Many years ago it was thought that the Universe's expansion rate was decreasing but, now we know that the Universe's expansion is accelerating.

The explanation I can offer is the following: matter formed at the time of the Big Bang, as well as, stars, planets and galaxies when they were closer to the center of the Universe; that is, closer to the point in time where the Big Bang occurred, was subject to greater attraction given the greater density of the basic attraction particles "BAP" that fill all the vacuum of the Universe.

It is true that vacuum in the Universe is also filled with the "simple basic particles" –"SBP" and, "basic repulsion particles" –"BRP". But there is a fundamental difference regarding the densities of these particles.

SBP's and BRP's fill the vacuum of the Universe with uniform density. Because of their repulsive nature they are constantly moving apart from each other, and therefore, they have a uniform density distribution. Hence, their density can be represented as a horizontal line (see next figure). On the other hand, the BAP's or dark matter has more density in the center of the Universe due to its attractive nature; and, its density can be represented as a curve.

"BAP" particles are not only more concentrated in the center of the Universe but, must also have a higher concentration within galaxies and around them, because they are attracted by the other BAPs that exists within the electrons and quarks that make up the stars and planets in every galaxy.

The topology and density of the "BAP" distribution is quite complicated to explain in this paper. I can only say that they are more concentrated in the center of the Universe and in the galaxies. Therefore, such concentrations can act as gravitational lens, many of them containing galaxies within. As we know many gravitational lenses have been discovered by the Hubble Telescope. Given this, we can determine where the Big Bang occurred due to the higher concentration of matter particles "BAP". It should be located at approximately 13.7 x 10^9 light years from us. If the Hubble telescope can ever find at around that distance, a gravitational lens without galaxies within it then, it could be said that it is the center of the Universe.

The topology of the "BRP" on the other hand, is very simple: they are distributed uniformly throughout the volume of the Universe.

It is generally assumed that all matter was created in the Big Bang. That matter was thrown away at the speed of light, and that as time passed, this matter was converted into stars, planets and galaxies. The rate of expansion of these stellar bodies, which form the visible Universe, was decreasing during a period due to the attractive force of the "BAP" or dark matter, which as I have said, is more

28

concentrated in the center of the Universe, and also because of the gravitational force of visible matter. So, we can say that the "BAP" decelerate the expansion velocity, but, that at present and in the future, "BRP" will be pushing all visible matter outwards thus, accelerating its expansion.

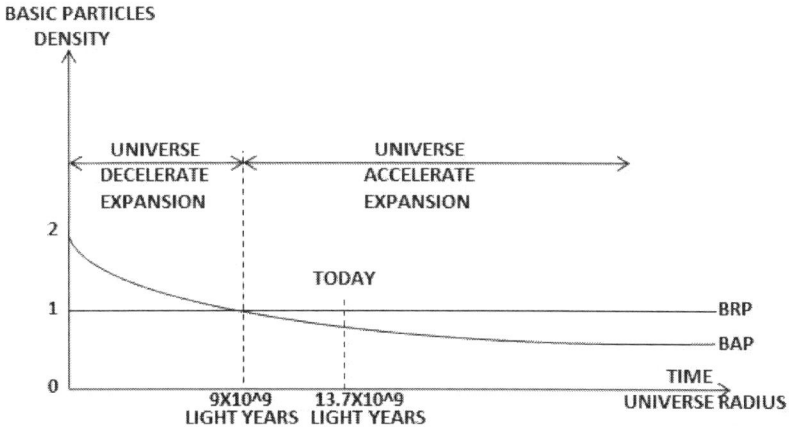

BASIC PARTICLES
DENSITY

| UNIVERSE DECELERATE EXPANSION | UNIVERSE ACCELERATE EXPANSION |

2

TODAY

1 ———————————————————————— BRP
————————————————————— BAP

0
TIME
9X10^9   13.7X10^9   UNIVERSE RADIUS
LIGHT YEARS LIGHT YEARS

There should be a moment that we could locate roughly where the radius of the visible Universe was nearly 9,000 million light years ago, and also, when the density of "BRP" or dark energy begins to be greater than the density of "BAP" or dark matter. Ever since that time, about 4,700 million light years, the expansion of the Universe started to accelerate.

My final conclusion is that in the future, visible matter will be much more under the effect of dark energy than under the effect of dark matter. And, that because of it the Universe will continue expanding in an accelerated mode.

29

16252061R00017

Printed in Great Britain
by Amazon